# Adapting to Climate Change
## 2.0 Enterprise Risk Management

## Dr Mark C. Trexler
Climate Risk Specialist, The Climatographers, Portland, OR, USA

## Laura H. Kosloff
Climate Law Specialist, The Climatographers, Portland, OR, USA

First published in 2013 by Dō Sustainability

87 Lonsdale Road, Oxford OX2 7ET, UK

ISBN 978-1-909293-46-5 (eBook-ePub)

ISBN 978-1-909293-47-2 (eBook-PDF)

ISBN 978-1-909293-45-8 (Paperback)

A catalogue record for this title is available from the British Library.

Dō Sustainability strives for net positive social and environmental impact. See our sustainability policy at **www.dosustainability.com**.

Page design and typesetting by Alison Rayner

Cover by Becky Chilcott

For further information on Dō Sustainability, visit our website: **www.dosustainability.com**

# DōShorts

Dō **Sustainability** is the publisher of **DōShorts**: short, high-value ebooks that distil sustainability best practice and business insights for busy, results-driven professionals. Each DōShort can be read in 90 minutes.

## New and forthcoming DōShorts – stay up to date

We publish 3 to 5 new DōShorts each month. The best way to keep up to date? Sign up to our short, monthly newsletter. Go to **www. dosustainability.com/newsletter**, or **click here to sign up to the Dō Newsletter**. Some of our latest and forthcoming titles include:

- *The Changing Profile of Corporate Climate Change Risk*
  Mark Trexler & Laura Kosloff

- *Promoting Sustainable Behaviour: A Practical Guide to What Works*   Adam Corner

- *Sustainable Transport Fuels Business Briefing*   David Thorpe

- *The First 100 Days: How to Plan, Prioritise & Build a Sustainable Organisation*   Anne Augustine

- *Full Product Transparency: Cutting the Fluff Out of Sustainability*
  Ramon Arratia

- *Making the Most of Standards: The Sustainability Professional's Guide*   Adrian Henriques

- *How to Account for Sustainability: A Business Guide to Measuring and Managing*   Laura Musikanski

- *Sustainability in the Public Sector: An Essential Briefing for Stakeholders*   Sonja Powell

- *Sustainability Reporting for SMEs: Competitive Advantage Through Transparency*   Elaine Cohen

- *REDD+ and Business Sustainability: A Guide to Reversing Deforestation for Forward Thinking Companies*   Brian McFarland
- *How Gamification Can Help Your Business Engage in Sustainability*   Paula Owen
- *Sustainable Energy Options for Business*   Philip Wolfe

## Subscriptions

In addition to individual sales of our ebooks, we now offer subscriptions. Access 60+ ebooks for the price of 6 with a personal subscription to our full e-library. Institutional subscriptions are also available for your staff or students. Visit **www.dosustainability.com/books/subscriptions** or email **veruschka@dosustainability.com**

## Write for us, or suggest a DōShort

Please visit **www.dosustainability.com** for our full publishing programme. If you don't find what you need, write for us! Or suggest a DōShort on our website. We look forward to hearing from you.

.........................................................................................................

# Abstract

**PUBLIC DISCUSSION OF ADAPTATION** and resilience to extreme events has picked up dramatically in the aftermath of Hurricane Sandy, which hit the Northeastern United States in late October 2012. Most companies do not yet recognize what it means to adapt to future climate change, and do not yet see it as a business priority. This DōShort tackles two key questions facing decision makers: 1) Is adaptation worth it to me? and 2) If it is worth it, can I really tackle it?

If a company has reason to worry about the potential impacts of weather on its operations and supply chains, it probably has cause to worry about climate change. Many companies believe they already incorporate climate change adaptation into corporate planning; in most cases, however, these companies are referring to efforts to manage conditions that they already are experiencing, rather than preparing for forecasted impacts of climate change. They are, in effect, "adapting to the weather." For many companies, focusing on today's weather and not tomorrow's climate leaves a lot of risk on the table, especially if the climate continues to change faster than many climate models have projected.

The uncertainties associated with forecasting climate change on a timeframe and at a scale that is relevant to corporate decision making can appear daunting. It is not necessary, however, to have perfect information to advance corporate preparedness for and resilience to climate change. Companies can improve their ability to make robust

# ABSTRACT

decisions under conditions of uncertainty without perfect information (such information not being likely to ever arrive anyway). A Bayesian approach to reducing uncertainty over time can cost-effectively support companies in understanding and managing many potential climate risks and can avoid the need to depend on future predictions. Instead, initial effort can focus on where a company will have confidence in its analysis and the ability to influence its level of risk, namely in assessing its exposure and vulnerability to climate hazards. As the hazards themselves become more clear, risk management strategies can be quickly adapted.

# About the Authors

**DR MARK C. TREXLER** has 25 years of experience advising companies on climate change risks. He joined the World Resources Institute in Washington, DC in 1988, working on the first carbon offset projects. He founded Trexler Climate + Energy Services (TC+ES) in 1991, the first consultancy to specialize in corporate climate risk management. TC+ES chalked up a lot of 'firsts', including many more of the early carbon offset projects, the first corporate greenhouse gas (GHG) inventories, the first corporate climate risk management strategies, and the first climate neutral products and services. Mark has worked with companies around the world, including many electric utility and oil and gas companies, and has served as a lead author for the Intergovernmental Panel on Climate Change. Mark was Director of Climate Risk for Det Norske Veritas from 2009 to 2012. Mark can be reached at **mark@climatographer.com**.

**LAURA KOSLOFF** has worked as an environmental attorney since 1985, including as Associate Editor of the Environmental Law Reporter in Washington, DC, as a trial attorney for the US Department of Justice, as General Counsel to Trexler Climate + Energy Services (TC+ES), and as Associate General Counsel to EcoSecurities Group plc. In her General Counsel roles she supported teams of more than 20 climate change consultants, managing all contract negotiations as well as wide-ranging HR and IP issues across offices in multiple countries. Laura was the first Chair of the American Bar Association's Climate Change Committee, and has served in numerous leadership capacities within the organization, most recently chairing the group's annual meeting. Laura can be reached at **laura@climatographer.com**.

# Contents

# CHAPTER 1

# Introduction

**PUBLIC DISCUSSION OF ADAPTATION** and resilience to extreme events has picked up dramatically in the aftermath of Hurricane Sandy, which hit the northeastern United States in late October 2012. The public discussion has also drawn in the business community, based on a slowly growing perception among business leaders that the weather is changing in ways that put business infrastructure and supply chains at risk. In October 2012, for example, the global trade association of oil and gas companies (IPIECA) held its first internal workshop to explore climate change adaptation efforts and needs in the sector.[1]

This DōShort is not an adaptation how-to manual. Beyond the fact that adaptation efforts will have to be highly context-specific, and the growing recognition by business of changing weather patterns, most companies do not yet recognize what it means to adapt to future climate change, nor do they see it as a business priority. Rather, this DōShort will address two key questions that corporate executives should be able to answer in the affirmative before they can prioritize a topic like climate change adaptation:[2]

- Is it worth it to me as a corporate decision-maker to tackle this topic?

- If it is worth it, do I have the ability to achieve my stated objectives, risk reduction or otherwise?

The first of these two questions encompasses the broader aspects of corporate climate change risks that we tackled in *The Changing Profile of Corporate Climate Change Risk* (Trexler and Kosloff, DōSustainability, Oxford, 2012). We introduced several ways that climate change is linked to business risk (and opportunity):

- *Physical risks*, including direct impacts of climate change on company operations, supply chains, and financial performance, among many other variables. Physical risk is based on existing and forecasted climate change; this DōShort focuses on physical risk.

- *Brand risks*, including the impact of consumer and stakeholder perceptions on corporate competitiveness. Brand risk (and perceived branding opportunities) has been a primary driver of corporate mitigation efforts to date.

- *Policy risks*, e.g. the impacts of climate policies and regulatory mandates on corporate competitiveness, and the susceptibility of business models to policy-driven variables like a price on carbon.

- *Structural market risks*, including changing supply and demand for company products and services in a carbon-constrained world, and responses to changing rates of technology innovation.

- *Liability risks*, including for retroactive or future greenhouse gas (GHG) emissions. We consider liability risks separately from policy risk because liability could also arise through litigation as well as from policy measures.

In our 2012 DōShort we also framed several climate change risk scenarios that help bring relative risk probabilities into the discussion.

In this new DōShort, we supplement that framing to include a rough estimate of the magnitude of adaptation needs associated with each scenario:

- *Scenario 1: Issue collapse.* The pressure for policy action, and the need to materially adapt to climate change, could come to an end if the current understanding of climate change science were to reverse course. The best available science suggests this is a very low probability scenario.

- *Scenario 2: Stay the (policy) course.* This scenario involves pursuing a wide range of policies and measures, but at a level too low to stabilize greenhouse gas (GHG) concentrations in the atmosphere. As a result, $CO_2$ concentrations continue to rise, reaching 700–900 parts per million (ppm) by the end of this century. The past 25 years of climate policy efforts suggest that this scenario constitutes the most likely 'Business as usual' scenario. Adaptation needs are high and grow dramatically over time.

- *Scenario 3: Technology-led transition to a low carbon economy.* In this scenario, we avoid dramatic climate change through accelerated rates of technology development and deployment, even in the absence of aggressive public policy (including material carbon pricing). Even technology optimists, however, emphasize the need for a supportive technology policy framework, including internalization of a material carbon price to support technology deployment decision-making. Thus, this is a low probability scenario, but it would substantially reduce needed adaptation as compared to the 'Stay the policy course' scenario.

- *Scenario 4: Policy-driven atmospheric stabilization.* In this scenario, we stabilize atmospheric $CO_2$ concentrations between 450 and 650 ppm (they are at 400 ppm today). The necessary political will would most likely result from public outrage associated with extreme climatic events, and triggering of a Climate Response Tipping Point (CRTP).[3] The probability of this scenario seems likely to increase over time as climate change manifestations become more obvious. The longer it takes to get to the CRTP however, the greater the anticipated adaptation needs.

- *Scenario 5: Policy-driven return to 350 ppm $CO_2$.* This scenario sees $CO_2$ concentrations reaching 450–550 ppm in the atmosphere, but we are then able to return concentration to 350 ppm through aggressive policy and technology interventions (350 ppm often being characterized as the level needed to protect the world's oceans from excessive acidification). Today's best available information suggests this is a very low probability scenario, with complicated implications for adaptation needs since it's hard to know how the climate would respond to this scenario.

Potential adaptation needs range widely across these scenarios. The most likely scenario ('Stay the policy course'), also results in the greatest magnitude of needed adaptation. We focus in this DōShort on this 'Business as usual' scenario, and ask how companies can best answer their 'is it worth it, can I do it' questions using the best available information, and based on accepted risk management principles.

While we focus in this DōShort on the physical risks of climate change, physical risks are not the only climate risks companies face. Indeed, the 'Stay the policy course' scenario could easily generate a 'risk double whammy', starting with climate change adaptation needs, followed by (when the Climate Response Tipping Point is reached) much more radical mitigation mandates than most companies have ever anticipated.

Given the current state of global climate policy, however, climate change itself may well be the more material near-term business risk for most companies and regions, requiring a growing risk management response. Our exploration of corporate climate change adaptation is broken into four sections:

1.  An overview of climate change risk and the adaptation response.

2.  Framing adaptation from a business perspective.

3.  The needs and barriers of an adaptation response.

4.  Developing an 'adaptive' enterprise risk management strategy.

In our experience, it is easier for corporate decision-makers to arrive at an affirmative answer to the 'is it worth it' question when it comes to climate risk than to the question of 'can I do it', particularly when it comes to climate change adaptation. The uncertainties associated with forecasting climate change on a timeframe and at a scale that is relevant to corporate decision-making can appear daunting. A key step in answering the 'can I do it' question, however, is to properly frame what 'it' is. Coming up with a low-risk high-certainty adaptation strategy where the results of cost–benefit analysis suggest clear courses of action for 'climate proofing' a company's assets and supply chain for decades into the future is simply unrealistic. We shouldn't let the perfect become the enemy of the good, given that reducing uncertainty and risk are achievable and desirable business objectives. After all, climate change is not the only topic for which we cannot predict the future with certainty (indeed, are there any?).

.....................................................................................................................

# CHAPTER 2

# Climate Change Adaptation: An Overview

**CLIMATE CHANGE MITIGATION** and climate change adaptation are generally considered distinct topics, mitigation being characterized as reducing climate change hazards, and adaptation as reducing one's exposure and vulnerability to those hazards.

Whether responding to public policy aimed at mitigating climate change, or to the physical impacts of climate change on their operations, companies have to adapt their business strategies to the risks and opportunities of the changing environment in which they operate. In principle, the concept of 'business adaptation' to climate change could encompass business responses to policy as well as physical risks. In this volume, however, we will use the term more narrowly, focusing on adaptation to the physical impacts of climate change.

Any useful discussion of adaptation to climate change requires that we start with a common understanding of how we are defining the term. For this DōShort, we turn to the Intergovernmental Panel on Climate Change which defines adaptation as 'initiatives and measures to reduce the vulnerability of natural and human systems against actual or expected climate change effects'.[4]

The need to adapt to climate change has been anticipated since climate change became a topic of political and policy discussion more than

20 years ago. In recent years, however, the relative amount of attention devoted to adaptation has soared given the:

- growing recognition that mitigation efforts have fallen dramatically short;

- steadily increasing damages from natural disasters;

- industrialized country commitments to cover adaptation costs in developing countries;

- improving capabilities when it comes to forecasting localized climate change.

Global adaptation initiatives now include the Loss and Damage Program, the Adaptation Fund, and National Adaptation Planning Activities (NAPAs), all being pursued under the umbrella of the United Nations Framework Convention on Climate Change (UNFCCC). Cities and municipalities around the world have launched climate change adaptation initiatives, including high profile efforts in London and Venice to stay ahead of sea level rise with massive engineering projects. International development banks like the Inter-American Development Bank and the Asian Development Bank increasingly call for 'climate proofing' of their development projects.

Losses from natural catastrophes in 2011 set a new record, estimated at $148 billion globally; approximately a third of this amount was insured.[5] Representative events are shown in Figure 1. The losses in 2011 continued a trend that insurance company Munich Re has documented for more than 30 years; during this time, damages from natural disasters have increased fivefold[6] (see Figure 2). Although climate change is certainly not the only variable contributing to escalating damage figures,

. . . . . . . . . . . . . . . . . . . . . . . . . . . . . . . . . . . . . . . . . . . . . . . . . . . . . . . . . . . . . . . . . . . . . . . .

## FIGURE 1. Significant weather-related loss events in 2011

**SOURCE:** D. Grossman, 2012. *Physical Risks from Climate Change* (Calvert Investments, Ceres, and Oxfam America). 'Significant Weather-Related Loss Events in 2011'. Adapted from: Munich Re, Geo Risks Research, NatCatSERVICE, 2012.

. . . . . . . . . . . . . . . . . . . . . . . . . . . . . . . . . . . . . . . . . . . . . . . . . . . . . . . . . . . . . . . . . . . . . . . .

Munich Re has concluded that upward trends in damage claims from extreme events can only be fully explained if one includes a growing climate change 'signal'.

As a result, many business players perceive climate change adaptation as a real-time need based on already changing climate conditions. In discussing weather extremes observed in Windsor, Canada, the city's environmental coordinator recently stated: 'Regardless of what the projections are, we're seeing it, so how do we start adapting to it?'[7]

Companies too are feeling the impacts of what many perceive to be changing weather, and more and more examples are being put forward to justify the business case for undertaking climate change

...........................................................................................

**FIGURE 2. Relative trend in natural loss events worldwide between 1980 and 2010.**

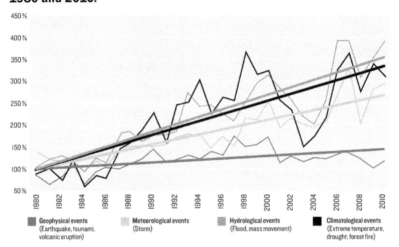

**SOURCE:** National Round Table on the Environment and the Economy, 2012. *Facing the Elements: Building Business Resilience* (Ottawa: National Round Table), at p. 85.

...........................................................................................

adaptation initiatives:[8]

- Under Armour®, maker of athletic apparel, registered elevated retail inventory levels for the 2011–2012 winter season as a result of 'unseasonably warm weather'.

- The Texas drought during the summer of 2011 forced utility Constellation Energy to buy extra power at peak prices, reducing third-quarter earnings by $0.16 a share.

- The Australian floods of 2010–2011 led to more than $2 billion in insurance claims, helping reduce Munich Re's fourth quarter profit by 38%.

- Australian weather extremes during 2011 reduced Rio Tinto's earnings by $250 million due to declines in production and weather-related shutdowns.

- During winter 2011–2012, total skier visits to Vail Resorts properties declined more than 15% due to lack of snow.

Another increasingly cited example is that of the 2011 floods in Thailand, which are estimated to have disrupted the operations of more than 14,500 companies globally.[9]

There is evidence to support that there is a 'new normal' when it comes to the changing climate. A 2012 analysis of trends in global temperature anomalies, for example, suggests that the anomalies' distribution is increasingly biased more toward hotter than colder anomalies. **Figure 3** illustrates the changing distribution around the mean of summer temperature anomalies in the Northern Hemisphere. To generate this visualization, temperature extremes for the period 1951–1980 were 'normalized' into a bell curve distribution (representative of a 'stable climate' in which low and high temperature extremes will be equally

**FIGURE 3. Shifting distribution of summer temperature anomalies.**

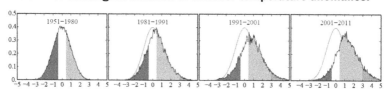

**SOURCE:** Hansen, J., Sato, M. and Ruedy, R. 2012. *Science Brief – The New Climate Dice: Public Perception of Climate Change*, August, available at: **http://www.giss.nasa. gov/research/briefs/hansen_17/**.

represented). Figure 3 suggests a strikingly different 'new normal' that is about one standard deviation 'warmer' than the prior normal, and increasingly favoring hot rather than cold anomalies.

## What will we have to adapt to?

One of the biggest questions facing climate change adaptation planning is the question of 'what will we have to adapt to?' If we were able to limit climate change to levels that avoid dangerous interference with existing climate systems (the stated goal of the UNFCCC), we could obviously bound the need for adaptation. It is increasingly clear, however, that the 2°C threshold commonly characterized as 'safe' under the UNFCCC is a political rather than scientific construct. Numerous studies document the occurrence of significant changes in climate, and yet average global temperatures have risen by less than 1°C. The last time global temperatures were 2°C higher than in 1900, global sea levels were 4–6 meters higher than today. Hundreds of millions of people live within that band of land currently above sea level; one could realistically question the societal 'safety' of such a level of change.

There probably is not much point to focus on the 'safe' 2°C threshold, however, as no policies currently exist to prevent average global temperatures from rising by more than 2°C in coming decades. As concluded in the 2012 report *Degrees of Risk*, the best case for the year 2100 is 2–3°C of average global temperature change; the worst case is 6–8°C (see Figure 4).

All of these temperature forecasts use the year 2100 as the 'end date' for analysis. Climate modelers have more confidence in models over longer modeling periods, and longer time horizons tend to deliver more

..........................................................................................

**FIGURE 4. Degrees of risk.**

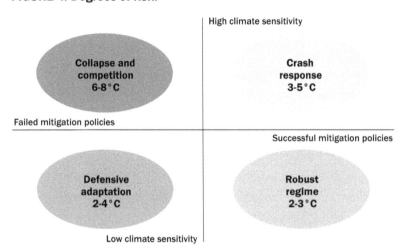

SOURCE: Mabey, N., et al. 2011. *Degrees of Risk: Defining a Risk Management Framework for Climate Security* (London: Third Generation Environmentalism Ltd). Figure 7.1: Climate Security Scenarios in 2100 based on a 2°C global mitigation target, p. 123.

..........................................................................................

draconian examples of the expected level of change. Based on the best available information, the implications of climate change in the year 2100 are almost certainly game-changing. One study characterized a 6°C increase in average global temperatures as likely to result in 'abrupt, run-away climate change' and 'mass loss of human life'.[10] Yet the year 2100 is well beyond the decision-making timeframes of virtually all public entities, much less private entities. No matter how dramatic, climate change in 2100 is just too far away to form the foundation of near-term societal or corporate risk-management discussions. Moreover, unfortunately, climate models are proving less useful than we might have hoped in forecasting the changes in extreme events that are the source of much of the current concern over climate change. If events are becoming more extreme and more frequent, it doesn't really matter that average global temperatures are changing more slowly.

Adaptation planning therefore has to be characterized by three variables that fundamentally differentiate it from most climate change forecasting. First, adaptation planning has to address the time horizons of relevance to very different types of organizations, from public entities willing to look a few decades into the future, to private entities that may not look more than a few years into the future. Second, adaptation planning has to address the local manifestations of global climate change. A 2°C change in average global temperatures, for example, will actually reflect a wide range of local temperature outcomes. In parts of California, a 2°C change at the global level is expected to result in 6–8°C locally. Third, adaptation planning has to incorporate outcomes that the climate models have difficulty forecasting.

Recent climate risk analyses are trying to focus more attention on nearer-term climate change impacts and adaptation needs in order to make the perceived impacts of climate change more 'real' for stakeholders and decision-makers. Sandia National Lab's *Assessing the Near-Term Risk of Climate Uncertainty: Interdependencies among the U.S. States* limits its forecasting horizon to 2050, arguably within the planning horizon of major infrastructure projects being planned today, and is still able to forecast dramatic reductions in the availability of cooling water for power plants (among many other impacts).[11] The World Bank's recently released *Turn Down the Heat: Why a 4 Degree Centigrade Warmer World Must Be Avoided*, tries to draw as detailed a picture of a 4°C warmer world as possible, again to make the implications of climate change seem more 'real' for today's risk management conversations.[12]

The direct physical manifestations of climate are of course just one aspect of 'what will we have to adapt to?' How the impacts of climate change will translate into economic impacts is also of paramount importance to adaptation planning. As one might imagine, however, it is not easy to robustly quantify the future economic implications of climate change. Prevailing economic wisdom assumes that climate change will affect global economic growth only modestly; however, that is far from a universally held view. Harvard's Jody Freeman suggests that conventional estimates dramatically understate the potential economic implications of climate change by failing to account for complex and cascading chains of economic impacts. **Table 1** compares conventional economic wisdom of a 0.5% reduction in future US Gross Domestic Product (GDP) resulting from 2°C of climate change, to Freeman's much higher estimate of 10–20%.

Seemingly modest changes in macro-economic indicators can also hide substantial diversity in how climate change could affect individual business sectors. Some sectors may even conclude that they will end up being better off. As Cameco, one of the world's largest uranium producers, concluded: 'overall the potential climate change benefits outweigh the potential drawbacks'; this led the company to not categorize climate change as a business risk.[13]

..............................................................................................

**TABLE 1. Quantitative adjustments to conventional estimates of climate change impacts.**

| Factors Considered | Conventional Estimates of Reduction in U.S. GDP (%) | Marginal Impact on Annual GDP (%) |
|---|---|---|
| Conventional IAM Estimate | 0.5 | 0.5 |
| Optimism About Temperature Rise | 0 | 1 |
| Asymmetry Around Point Estimates | 0 | 0.5 |
| Catastrophic Events | 0 | 0.5–3 |
| Nonmarket Costs | 0 | 1.4–3.5 |
| Export Losses | 0 | 1.5 |
| SUBTOTAL | 0.5 | 5.4–10 |
| Growth and Productivity | 0 | Double Above Impacts |
| TOTAL | 0.5 | 10.8–20 |

**SOURCE:** Freeman, J. and Guzman, A. 2009. Climate change and U.S. interests. *Columbia Law Review* 109:1531–1602, at p. 1596, Table 3: Quantitative Adjustments to Conventional Estimates of Climate Change Impacts.

..............................................................................................

Further complicating discussions of climate change adaptation is that estimating the costs of adaptation is a step removed from estimating the costs of climate change. Estimating future adaptation economics depends not only on climate change forecasts, compounded by uncertainties regarding economic impacts and vulnerabilities, but on choices made as to what adaptation measures to undertake. The costs of adaptation, therefore, are not necessarily a function of either the costs of climate change or the costs of mitigating climate change, and such estimates need to be treated with care. Near-term estimates of global adaptation costs range from $40 to $140 billion per year, and few estimates extend even out to 2030 (see Figure 5). Yet very little of the climate impacts literature focuses on near-term impacts.

Overall, there is no simple or particularly robust answer to the question of 'what will we have to adapt to, by when, and what will it cost us', both

**FIGURE 5. The costs of adaptation are uncertain: estimates vary substantially and are incomplete.**

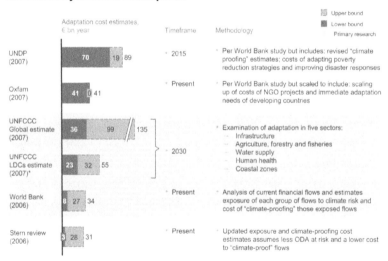

**SOURCE:** ClimateWorks Foundation, Global Environment Facility, European Commission, McKinsey & Company, The Rockefeller Foundation, Standard Chartered Bank and Swiss Re. 2009. *Shaping Climate Resilient Development: A Framework for Decision Making*, a report of the Economics of Climate Adaptation Work Group, at p. 124, 'The costs of adaptation are uncertain: estimates vary substantially and are incomplete.'

because of uncertainties over how much the climate change will change and how fast, and because the economic exposure and vulnerability of societies and business sectors to those changes range so widely.

Does this suggest that risk-based conversations can simply be postponed? The Sandia National Labs report introduced above addresses the role of uncertainty in risk-based planning, noting that the level of uncertainty in climate forecasting itself creates substantial business risk. The study concludes that: 'It is the uncertainty associated with climate change that validates the need to act protectively and proactively.'[14]

## Are there limits to adaptation?

Rex Tillerson, the CEO of Exxon, recently acknowledged the reality of climate change and Exxon's contribution to climate change, but suggested that the need of developing countries for fossil fuels trumps climate change as a priority. Mr Tillerson concluded: 'We have spent our entire existence adapting. So we will adapt to this.'[15] Average global temperatures during the last 10,000 years have fluctuated within a remarkably narrow band (about 1°C). To suggest that because human societies have been able to adapt to a combined 1°C of change over the last 10,000 years, we will be able to adapt to 6–8°C in the next 100 years is not a risk-based perspective. This raises the obvious question: are there limits to our ability to adapt to climate change, especially when we have more fully exploited the carrying capacity of the planet today than at any point in the past (see Box 1).

## BOX 1. Past performance is no guarantee of future results

Notwithstanding frequent examples to the contrary, we naturally tend to assume that the future will be an extension of the past. The 20th century was a remarkable period for human and economic development, and we tend to assume that the 21st century will reflect 'more of the same'. McKinsey's 2010 Resource Revolution report challenges this conventional wisdom, in particular documenting that it's just not mathematically feasible to project that key trends will continue into the future. The report notes that the dramatic natural resource price reductions between 1900 and 2000 were wiped out between 2000 and 2010, with real commodity prices rising 147%, and the price of drilling an oil well rising by 100% over the same period. More than 40 million people were driven into poverty by rising food prices in 2010 alone. Looking forward, 3 billion new consumers will enter the global middle class by 2030, accompanied by an increase in the demand for steel of 80%. As shown in the figure below, the rate of increase in key resources between 1990 and 2010 and 2010 and 2030 would have to increase by 32% for primary energy, by 139% for water, and as much as 249% for land, to maintain historical trends.

Although McKinsey also identifies a wide range of opportunities to address projected resource shortfalls, the bottom line conclusion of the report is that past performance is no guarantee of future results when it comes to the availability of key natural resources. Many of the challenges cited by McKinsey will only be aggravated by a changing climate, most obvious among them water and land availability. Growing and more affluent human populations will need much more water and land at exactly the same time that these resources are being stressed by climate change.

**Additional supply would have to accelerate by up to 250 percent versus the past 20 years in a supply expansion case**
Additional supply needed over 20-year time frame[1]

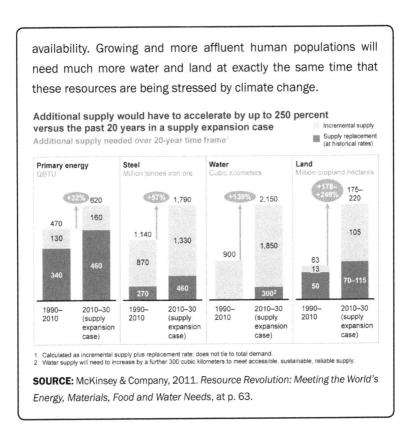

1 Calculated as incremental supply plus replacement rate; does not tie to total demand.
2 Water supply will need to increase by a further 300 cubic kilometers to meet accessible, sustainable, reliable supply.

**SOURCE:** McKinsey & Company, 2011. *Resource Revolution: Meeting the World's Energy, Materials, Food and Water Needs*, at p. 63.

What are the limits to adaptation? We don't really know. A curious aspect of the adaptation issue, however, is that whoever might ask the question in the future, simply by virtue of the fact that they are in a position to ask the question, is likely to conclude 'yes, we adapted'. But they won't be speaking for either the people or the natural systems that didn't fare so well. Defining the 'success' of societal adaptation efforts, or identifying societal limits to adaptation, is a challenging topic that has yet to be widely tackled.

# CHAPTER 3

# Setting the Business Stage for Adaptation

**BUSINESS DISCUSSIONS OF CLIMATE RISK** have focused historically on policy and regulatory risk; corporate climate risk management has focused on influencing policy and on reducing corporate carbon footprints through measures ranging from energy efficiency to carbon offsets and renewable energy credits. Recognition of the physical risks of climate change is increasing, however. Business discussions are proliferating on how climate change already is disrupting corporate supply chains. The specter was raised in the aftermath of Hurricane Sandy that companies could be held liable for their lack of resilience to extreme events, and that companies failing to adequately characterize climate risks could run afoul of new Securities and Exchange Commission reporting guidance in the United States. It is increasingly common to see climate change impacts and adaptation issues raised in environmental impact analyses and in permitting and siting processes, something that was almost unheard of just a few years ago. In practice, adapting to climate change will involve huge numbers of both corporate and government decisions, engineering and otherwise.

Of course, policy will have a major role in influencing adaptation. As with mitigation, many market barriers can impede a societally optimal level of adaptation, and public policy can help overcome those barriers. A key priority will be to avoid mal-adaptation. When private insurers pulled out of some insurance markets along the Florida coast, for example, the State of Florida created a public insurance pool to provide the missing coverage. In a particularly high-risk example of mal-adaptation, the state's action encouraged building and re-building in highly vulnerable areas, and likely puts Florida taxpayers on the hook for the virtually inevitable bankruptcy of such an insurance pool in the face of future extreme events. Another example of mal-adaptation can be seen in the recent situation in North Carolina where legislators prohibited the use of new sea level rise forecasts because of a fear that their use in land-use planning would undercut property prices.[16]

# Climate change mitigation versus climate change adaptation

Twenty-five years ago AES was the first company to openly pursue a corporate climate change mitigation strategy by linking carbon offset projects to power plants it was building in the United States. Many companies followed AES down the carbon offset path or chose mitigation paths of their own (including energy efficiency, renewable energy, and consumer education among other vehicles). A variety of mitigation 'business cases' encouraged these efforts, from needing to 'be at the policy table', to actively reducing corporate risk through reducing corporate emissions. As companies tried to develop and implement voluntary climate change mitigation strategies, however, they found that many of these business cases seemed to look at the world through rose-colored glasses. When they went to apply those business cases to their own situations, companies found that:

- While energy efficiency and other measures resulted in some direct financial savings, reductions were often relatively modest.

- Voluntary corporate programs were limited in potential scope; thus, no matter how aggressive a company's mitigation targets, the physical risks they faced from climate change wouldn't change.

- The link between voluntary mitigation programs and the ability to reduce policy and regulatory risk seemed tenuous. Indeed, in the absence of baseline protection policies, companies worried that their efforts to voluntarily reduce emissions might come back to haunt them through stricter reduction requirements when mandates did materialize.

- Consumers didn't understand climate change mitigation; instead of always enhancing brand image, voluntary mitigation efforts could easily lead to 'greenwashing' charges.

- Regulators (where involved) often weren't impressed by voluntary mitigation efforts, and wouldn't allow the costs of mitigation efforts to be passed through to ratepayers (in the case of electric utilities).

These factors, along with the continuing absence of climate policy, help explain why many companies have migrated away from climate change mitigation programs in recent years. Today, even as policy-makers and others call on the private sector to step in and substitute for failed societal efforts toward mitigating climate change, companies are not playing along.

It would be natural for companies to approach the climate change adaptation conversation skeptically given their perceived experience with climate change mitigation. Should they lump the two together in this way? No. There are key differences between the business attributes of climate change mitigation and adaptation that are relevant to developing a 'business case' for action:

- Unlike mitigation, 'selling adaptation' does not require convincing corporate decision-makers that material climate change policies (and a material carbon price) are right around the corner.

- The physical climate risks driving adaptation can involve the types of events (e.g. extreme weather) and outcomes (e.g. supply chain disruptions, corporate crisis management) with which decision-makers and investors are already familiar. This makes it easier to link adaptation to personal and corporate experiences.

- Adaptation is local while a lot of mitigation, particularly mitigation involving carbon offsets, is global and harder to explain to stakeholders.

- With adaptation measures you can be relatively confident that you'll reap the benefits of your own actions. The emissions reductions of many well-intentioned mitigation efforts, on the other hand, can be completely negated by actions outside your control (e.g. someone cutting down the forest you were trying to save).

- The psychology of adaptation triggers thinking about 'good' outcomes (e.g. resilience, preparedness), while mitigation tends to trigger thinking about 'bad' outcomes (e.g. higher energy prices, mandated cuts).

These differences have implications for how companies will perceive the priority of adaptation to climate change. These examples suggest that climate change adaptation may be easier for companies to grasp and prioritize. Other variables make the picture less one-sided:

- Like mitigation, adaptation can seem like a 'future problem' to companies focused on quarterly performance reports.

- Many companies were able to pursue inexpensive climate change mitigation strategies, e.g. through profitable energy efficiency or the purchase of cheap carbon offsets. Adaptation efforts could be more costly and with less immediate payoff.

- Some companies have derived significant brand benefit from their mitigation efforts, e.g. through initiatives like rainforest conservation or footprint reductions; this may be much harder to do with adaptation measures that will often have a lower public profile and less public relations benefit.

Overall, adaptation should be an easier subject for companies to get their heads around than mitigation, but this is by no means assured, at least in the short term when the physical risks of climate change are still gathering steam.

## The growing business recognition of adaptation as a need

Successful companies adapt to changes in their business and risk environments. What can we say about how business environments are changing as a result of the physicals risks from climate change?

- We can say with high confidence that the climate change *Hazards* facing business are increasing as atmospheric GHG levels increase. We can also reasonably anticipate the emergence of climate change Black Swans (e.g. unanticipated outcomes of major business significance) as Hazards increase.

- We can say with high confidence that business *Exposure* to climate change impacts is increasing. Economic activity continues to congregate along the world's coasts and waterways, where extreme events are often most obvious, even as sea level rise is accelerating and floods increase in severity. Exposure to climate change impacts is also increasing in many other areas, from global trade to public health.

- We can say with high confidence that corporate *Vulnerability* to climate hazards is increasing as exposed infrastructure becomes more costly to replace, as supply chains become more complex and more vulnerable to disruption given the rapid growth of 'just in time' distribution systems, and as the resource and other constraints increasingly stress the flexibility of the global economy (see **Box 1**). As already noted, the Thai floods of late 2011 are estimated to have materially affected operations at some 15,000 companies around the world, a powerful message of how vulnerable companies have become to weather events around the world.

Some business leaders express increasing concern over the growing level of risk based on increasing climate hazards, exposure and vulnerability. Recent corporate recognitions of such risk include:

- Coca Cola: 'We recognize that climate change has the potential to significantly affect the sustainability of our business and supply chain.'[17]

- Munich Re: 'Climate change affects the fundamentals of doing business, both yours and ours.'[18]

- Rio Tinto Alcan: 'Demonstrating leadership in climate change strengthens our long-term competitiveness.'[19]

Not surprisingly, we now more commonly see business-oriented publications targeting business climate risk and adaptation, including:

- ICF, 2008. *Adapting to Climate Change: A Business Approach.*

- PWC, 2010. *Business Leadership on Climate Change Adaptation.*

- National Roundtable on the Environment and the Economy, 2012. *A Primer for Business Executives on Adaptation.*

- International Business Leaders Forum, 2012. *The Business of Adapting to Climate Change: A Call to Action.*

- Partnership for Resilience and Environmental Preparedness, 2012. *Value Chain Climate Resilience – A Guide to Managing Climate Impacts in Companies and Communities.*

- Grossman, 2012. *Physical Risks from Climate Change – A Guide for Companies and Investors on Disclosure and Management of Climate Impacts.*

PWC's *Business Leadership on Climate Change Adaptation* explores rationales for undertaking adaptation efforts, from lower insurance rates and better employee health to improvement in a company's reputation and market competitiveness and justification of companies' 'license to operate' (see Table 2). Many of these rationales resemble those given for corporate mitigation efforts a decade or more ago. Many of these reports also make extensive use of anecdotes, often company-specific, in building a generic business case for adaptation.

**TABLE 2. 2010 PWC business leadership.**

| IMPACT AREA | RATIONALE FOR ACTION |
|---|---|
| New market and products | • Mapping climate impacts can help ensure business plans and investment and loan decisions make the most of climate risks and opportunities |
| Existing market and products | • Early response to changes in existing markets and products as a result of climate risks and opportunities could maintain or generate competitive advantage over peers |
| Security of inputs and supply | • Supporting suppliers to become climate resilient can secure raw material supplies and therefore production. Strategies on diversification of supplies can help spread the risks of supply chain disruptions |
| Cost of inputs and supply | • Awareness of how suppliers are affected can enable business to source from lower cost regions |
| Operations (continuity and costs) | • Early recognition of climate risks can help identify impacts and develop more effective business continuity strategies<br>• Understanding how climate risks could impact operational effectiveness can help decision making on investments that can manage or lower longer term operational costs |
| Fixed assets/ Infrastructure | • Incorporating climate risks into site selection can help maintain operational effectiveness and desirability of business locations<br>• Incorporating climate scenarios into the asset design and specification can reduce avoidable future expenses e.g. in retrofitting |

| | |
|---|---|
| **Regulation** | • Taking proactive steps to adapt to climate change can reduce the compliance or regulatory cost burden |
| **Licence to operate** | • Having a track record of assisting communities local to operations to adapt to climate impacts can support the social license to operate |
| **Insurance** | • Demonstrable management of climate risks can improve insurability and reduce the cost of premiums and claims |
| **Workforce** | • Helping future-proof the local community and employee conditions can ensure a mobile, healthy and safe workforce that can continue to operate effectively |
| **Local community** | • Helping future-proof the local community can improve reputations and support for the business |
| **Reputation** | • Disclosing how the business is managing climate risks and maximising opportunities can provide confidence to investors, consumers and other stakeholders<br>• Assisting communities to adapt and being seen as a solutions provider can provide reputational value |

**SOURCE:** PWC, 2010. *Business Leadership on Climate Change Adaptation: Encouraging Engagement and Action.* In collaboration with UNFCCC.

In reality, business adaptation needs and opportunities will vary substantially by business sector, by geography, by business model, and by individual firms' tolerance for different kinds of risk. Indeed, the business case for material adaptation efforts is likely to be almost company-specific at least in the short term.

Relatively few in-depth adaptation business cases yet exist. One is the analysis undertaken by the US utility Entergy after suffering $2 billion in damage from Hurricanes Katrina (2005) and Rita (2005).[20] Entergy's service territory is located along the Gulf Coast and includes extensive oil and gas industry infrastructure. Based on the study, Entergy anticipates that annual weather impacts to oil and gas infrastructure in the region will increase by an average of US $14 billion per year under its climate change scenario (see Figure 6). Entergy's analysis also suggests that prudent adaptation measures could keep the absolute level of weather-related risk close to today's level.

In California, the California Energy Commission has been studying the

**FIGURE 6. The exposure of business to climate change risk and opportunity: a rationale for action.**

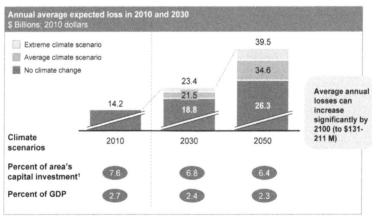

1 No climate change; includes impact of subsidence
2 Based on BEA historical average of capital investment (private and total government expenditures) as a percentage of GDP

**SOURCE:** Entergy. 2011. *Building a Resilient Energy Gulf Coast*, figure 3 at p. 7.

impacts of climate change for some time. In the case of electric utilities, for example, these impacts include:

- the impact of sea level rise on coastal energy infrastructure;

- the impact of higher temperatures on power plant operating efficiency;

- the impact of inadequate cooling water on the ability to operate fossil-fueled power plants; and

- the growing risk of fire to thousands of miles of transmission lines.[21]

Canada's Hydro Quebec, responding to an ongoing series of weather disruptions including low-stream-flow events, undertook detailed forecasting of water flows across its system using several climate scenarios. The utility concluded that changes to precipitation and snow patterns will materially impact its hydro-based generation system. Without adaptation, Hydro Quebec concluded, it might lose 14% of its net annual electricity output. Its analysis also concluded that adapting its system to anticipated changes might allow the utility to actually increase net power generation by as much as 15%.[22]

Some of the potentially most material aspects of near-term business adaptation may not even jump to mind in the context of business risk discussions. Changing summer ice conditions in the Arctic, for example, have already unleashed an explosion of economic activity relating to mineral and oil and gas development. With the Arctic Ocean potentially becoming ice-free during summer months in just a few years, the oil and gas industry anticipates new exploration opportunities. The shipping industry is paying close attention to the possibility of shaving thousands of miles off shipping routes by taking advantage of newly opening shipping channels that in the past could only be traversed by ice-breakers. These changes reflect adaptations to climate change, and perhaps help explain the characterization of the business response to climate change by one industry roundtable as: 'This is not just about coping with climate change, but prospering through it.'[23]

While many in the private sector highlight the 'business opportunity' of climate change wherever possible, business trade associations, investor associations, and consultants also characterize climate change adaptation as business risk management. Even now, however, as Harvard professor Jody Freeman argued with respect to the economic impacts of climate change, few companies have yet begun to recognize all of the ways that climate change impacts could directly and indirectly cascade through their specific corporate supply chains and operations.

CHAPTER 4

# A Closer Look at Business Adaptation Challenges and Needs

**MANY EXECUTIVES FACING 'CLIMATE MITIGATION FATIGUE'** will be naturally skeptical of similar-sounding adaptation business cases. It may sound good, it may look good, but can adaptation planning really deliver value?

The two questions previously introduced lie at the root of this question, and of any change in personal or corporate behavior:

- Is it worth it? Is it worth it to me as a corporate decision-maker to tackle this topic?

- If it is, can I do it? Can I successfully reduce perceived personal and/or corporate risk, or enhance opportunities, by implementing a climate change adaptation strategy?

What does it mean to be 'worth it' to a decision-maker to tackle climate change adaptation? Clearly the desires of stakeholders including customers are relevant. But in most cases, it will be difficult to argue that an adaptation response is worth it without acceptance that climate change poses material risk to a business in the first place. Only then will it make sense to ask the 'can I do it' question of whether identified risks can be significantly mitigated through a corporate adaptation process and strategy.

# Barriers to corporate climate change adaptation

Managing risk often has little to do with the nature of the risk itself, and more to do with the perceived self-interest behind alternative risk management strategies. Risk management self-interest, in turn, is influenced by a complex set of cognitive variables and biases.[24] It is easy to see how perceptions of self-interest can lead to the low priority often given to corporate climate risk. It is common to hear from electric utility executives, for example, that:

- 'If and when climate change becomes really material to us, policy-makers will realize that we (e.g. electric utilities) are too important to fail, so we'll be made whole by ratepayers or taxpayers. That being the case, spending a lot of time now worrying about climate risk and adaptation isn't likely to benefit shareholders, and could even penalize them if we're later told that because of our preparations we don't qualify for future assistance given to everyone else.'

- 'If something really bad were to happen in terms of climate change, something that we couldn't be made whole from, our financial and regulatory decision-making constraints probably wouldn't have allowed us to adequately hedge that risk anyway. And our competitors will also be in a similarly bad position, so it's unlikely to undermine our long-term competitive position.'

It's easy to understand the conclusion that if we expect a regulatory or societal helping hand to be ready when bad things happen in the future, incurring near-term costs to pro-actively manage the risk may deliver little relative benefit. This assumption flies in the face of current experience in other countries that have been grappling with adaptation questions. In the UK, for example, the government has said that it will not be responsible for individuals' and businesses' costs associated with future climate change impacts and adaptation needs.[25]

What's even more intriguing is a business perception that even if they cannot count on a helping hand from ratepayers or taxpayers when really bad things happen, pro-active risk management today may still not be a priority because their competitors won't have prepared either, and the competitive landscape will remain level. Yet it's a pretty big assumption that none of a company's competitors will have acted to enhance their competitive advantage under these circumstances, given that a lot of adaptation measures may not be easy to see. Nevertheless, managing risk is rarely free, and aggressive risk management can look pretty costly against other near-term funding priorities; it is not hard to understand how perceived corporate and decision-maker self-interest leads to these conclusions.

Thus, the business logic of any given risk management response cannot be interpreted solely through the lens of the hazard itself. A lack of corporate action on climate change adaptation most likely reflects self-interest as perceived through the filter of decision-maker incentives and perceptions of risk. If decision-making incentives discourage risk management, just as current flood insurance programs can incentivize rebuilding (sometimes repeatedly) in flood-prone areas, then it shouldn't be surprising that decision-makers act on adaptation in ways that appear suboptimal from a societal perspective.

Business decision-makers are also not immune to the psychological variables that generally color our perception and prioritization of risks. Climate change hazards, for example, don't tend to trigger our evolutionary fight or flight response, a dominant aspect of human risk management. We've personally never seen large-scale climate change before, leading us to discount its likelihood. And as humans we don't think well about probabilistic risks, making it difficult for us to deal with the fact that we can't confidently distinguish between climate change and natural weather variability. As John Celona stated, 'the human brain evolved to deal with lions and tigers on the Serengeti', emphasizing the importance of previously observed patterns to our perceptions of risk.[26] The psychology of human risk management is likely to be a more daunting barrier to companies' understanding and prioritizing their climate risks than is generally recognized.

# Adapting to the weather or to climate change?

Companies say they are devoting more and more attention to business initiatives aimed at climate change adaptation. In the first voluntary reporting of climate hazards by US insurance companies, over 75% of insurers said they anticipate increased natural hazards.[27] Recent reports discuss many anecdotal examples of corporate adaptation initiatives:

- Adding ski lift capacity at higher elevations to allow skiers to follow the snow up the mountain (Whistler Blackcomb, Canada).

- Replacing electric transmission poles with poles that can sustain higher speed hurricane winds (Entergy, USA).

- Planting more expensive Douglas fir seedlings (rather than pine) to increase forest resilience in the face of rising temperatures (Tolko, Canada).

- Focusing tree genetic work on adapting growing stock to future conditions, as well as integrating more pest-resistant species into the forest landscapes to slow down future pest outbreaks (J.D. Irving, Canada).

- Assessing future conditions in coffee-sourcing regions in order to help the company to work with local formers to improve crop resilience, or, where necessary, move to alternative crops (Green Mountain Coffee Roasters).

- Identifying populations most vulnerable to extreme weather disruptions due to having only one source of water, and establishing diversified sources to increase resilience (Anglian Water, UK).

- Diversifying vineyard holdings to have access to grapes from areas with cooler growing temperatures (Brown Brothers Wineries, Tasmania, Australia).

- Preparing for production of more fire-fighting aircraft (Bombardier Aircraft).

- Preparing for a rising demand for desalination plants to provide potable and irrigation water in a warmer world (SNC-Lavalin).

Distinguishing between adaptation-specific measures and 'business as usual' measures is not easy. Many business activities might be characterized as adaptation, but would have occurred even in the absence of perceived changes in the weather or forecasted climate change. Agriculture and forestry companies constantly strive to select for better genotypes, for example. It is easy to characterize much of that work as oriented toward climate change adaptation, but much of it would have occurred anyway.

......................................................................................................

**FIGURE 7. Yes or No: Is climate change incorporated within your organization's policies?**

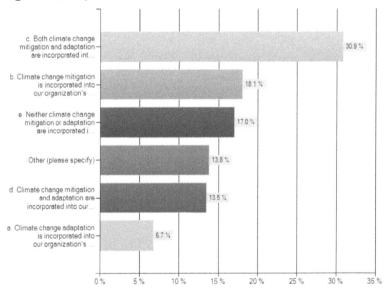

**SOURCE:** Association of Climate Change Officers, 2011. *Barriers to Climate Change Adaptation Plans: A Survey of Climate Professionals*, at p. 9.

......................................................................................................

How prevalent active climate change adaptation efforts are across companies and sectors is open to interpretation. In a recent survey, 90% of companies worldwide reported that they faced climate related impacts in the next three years, but only 30% were actively responding to those threats.[28] On the other hand, a 2011 survey by the Association of Climate Change Officers (ACCO) concluded that more than 50% of companies already incorporate climate change adaptation into corporate planning[29] (see Figure 7).

This conclusion is puzzling, given how recently the business community has started to discuss adaptation in any serious way. The authors believe that something else might explain the ACCO survey results. We developed and circulated our own survey to test this hypothesis. We circulated the survey to a small number of companies and their advisors (particularly attorneys providing strategic advice to major corporations). Although our sample was much smaller than ACCO's, corporate advisors were generally surprised to hear that their clients might already be adapting to forecasted climate change given their perception that most companies perceive climate risk (and adaptation needs) to be outside their planning horizon and companies simply don't know enough about what they would be adapting to. In response to the question 'what would it take to get companies to engage on adaptation?', advisors answered: 1) pressure from a major customer focusing on supply chain resilience; or 2) an improved understanding of the risks companies would be trying to manage through adaptation measures.

We don't dispute that 50% of the respondents to the ACCO survey reported that their companies are incorporating adaptation needs into their corporate strategies. We suspect, however, that most of these companies are mischaracterizing the nature of their corporate efforts.

When companies report they are already adapting to climate change, most of them are actually talking about efforts to grapple with weather events that they are already witnessing, rather than preparing for the forecasted impacts of climate change years into the future. They are, in effect, 'adapting to the weather'.

The fact that today's changing weather patterns are almost certainly, at least in part, an early manifestation of climate change does complicate our distinction between 'adapting to the weather' and 'adapting to climate change'. But we see the distinction as very real when it comes to risk management. Adapting to the weather is a risk-management 'no brainer'. Adapting to forecasted climate change, however, implies an acceptance of climate science, a longer-term planning perspective, and an understanding of the likely implications of future climate change for a company's operations.

If a company has reason to worry about changing weather and changing extremes, the company probably has cause to worry about climate change. Focusing on today's weather, and not tomorrow's climate, leaves a lot of risk on the table, especially if the climate is changing faster than climate models have projected. Treating adaptation as a 'rearview mirror' exercise, rather than as an opportunity to anticipate what will pop up in your headlights as you drive down the road, means you're likely to hit more obstacles in the road, and sustain more damage to your vehicle than necessary. Expressed in business terms, you're likely to be left with more stranded assets and missed business opportunities.

When pressed, many corporate planners don't disagree with our conclusion. They give several reasons for focusing on today's weather rather than tomorrow's climate:

- It's organizationally easier to respond to observed events rather than forecasted outcomes.

- Responding to observed changes in business conditions is often an operational responsibility, avoiding the need for difficult to procure strategic direction and leadership.

- Focusing on already observed changes to the weather makes it possible to avoid the often politically and organizationally contentious topic of 'what's causing climate change?'.

- We just don't know enough about climate change and its potential impacts to propose and implement real climate change adaptation strategies.

We don't question that companies should be prepared for the impacts and extremes they're already observing. Additionally, we recognize that 'adapting to the weather' can be characterized as a 'no regrets' approach to climate change adaptation; it's hard to go wrong by adapting to climate change you're already seeing. But 'no regrets adaptation' makes no more sense than 'no regrets mitigation' if it leaves a lot of a company's potential climate risk on the table. The key question facing decision-makers is whether they can realistically do better.

## What does adapting to climate change require?

Climate change adaptation is about reducing a company's future vulnerability to the impacts of climate change on a particular facility, its global operations, or its supply chain. As introduced earlier, before changing behavior to actively engage in climate change adaptation, corporate decision-makers need to answer the 'can I do it?' question.

## A CLOSER LOOK AT BUSINESS ADAPTATION
## CHALLENGES AND NEEDS

Incorporating climate change adaptation into corporate risk management strategies implicitly requires:

1. Understanding the nature of the climate hazards faced by the company. How do environmental conditions and extremes seem to be changing, and how do those changes manifest themselves on corporate operations? How might these conditions continue to change over the period of relevance to business operations and infrastructure?

2. Identifying and assessing corporate exposure to the identified hazards over whatever timeframe is considered relevant to a company's business model, corporate philosophy, and stakeholders.

3. Assessing economic vulnerability to the range of exposures identified, including operational and supply chain disruptions associated with changing weather and a changing climate.

4. Quantifying business risks based on the considerations of hazard, exposure, and vulnerability, yielding a prioritized list of risks to carry forward into the risk management process.

5. Identifying risk reduction opportunities, particularly for high-priority risks, and evaluating costs and benefits. How much will an adaptation strategy and implementation of adaptation measures cost? Risk reduction measures might be binary, e.g. one major company's decision to make no new infrastructure investments along the US Gulf Coast. But they can also be incremental, e.g. incorporating 'optionality' into infrastructure designs so as to be able to more easily retrofit the structure for increased resilience at a future date. In principle, companies should pursue those measures where

'marginal costs avoided' exceed the 'marginal cost of adaptation'. Ideally, companies should start with measures that maximize the risk reduction per unit of adaption expenditure.

6. Establish a process for reviewing and updating the assumptions and conclusions of the five previous steps. The National Center for Atmospheric Research in Boulder, CO, for example, has embarked on a project to model future climate across the continental United States using 4 km grids, a major advance over today's model resolutions that will justify a fundamental revisiting of many adaptation strategies. This is just one example of the kinds of information that companies should incorporate into adaptation risk management plans.

There are two unfortunate facts that quickly come to the fore when companies evaluate this list:

1. Individual companies have almost no ability to mitigate the climate hazards they face, even if they can often influence their exposure and vulnerability to those hazards.

2. Even if companies can reasonably understand, quantify, and even manage their exposure and vulnerability to climate hazard(s), the hazards themselves may remain subject to substantial uncertainty, e.g.

   • The full range of ways in which climate change will manifest itself across company operations, supply chains, and infrastructure plans.

   • The speed with which such changes will manifest themselves and their business materiality.

- Changes in investor, customer, and stakeholder behaviors and expectations based on the manifestations of climate hazards.

Not only can't we predict these outcomes, the uncertainties won't disappear. Even as the future is revealed to us through climate change itself, and as our ability to forecast future climate change improves, we will always be dealing with probability distributions around business-relevant outcomes, and changes in those probability distributions over time. The final assessment of climate risk is likely to be characterized by substantial uncertainty, and it's useful for decision-makers to understand the basics of such uncertainty.

## The basics of forecasting climate change

The starting point for most climate forecasts are general circulation models (also termed global climate models or GCMs). These models have become much more reliable in their ability to recreate historic and current climate, and scientists are increasingly confident in their ability to forecast future climate, particularly over longer time periods and at larger geographic scales.

GCM results incorporate three well understood uncertainties that are important to adaptation planning:

- *Natural variability*. Although our improved understanding of weather and our improved ability to forecast the weather con-stitutes one of the great scientific successes of the 20th century, little about the weather can be accurately predicted more than five to seven days ahead. This is the result of the 'chaotic' nature of weather systems, inherently limiting the ability of GCMs to precisely simulate future conditions no matter how much modeling power is brought to bear.

- *Scenario uncertainty.* Many factors are involved in forecasting future GHG and aerosol levels in the atmosphere, both important contributors to future climate change. Some of these factors relate directly to human activities, including population growth, economic activity and political and policy decisions. Others are natural variables (e.g. the rate at which oceans and forests will take up or release carbon dioxide in the future). These uncertainties manifest themselves in scenarios with a wider and wider range of potential outcomes the farther we look into the future.

- *Model limitations.* Any climate model has to make simplifications about how heat, water, and energy are added to or transferred between the atmosphere, the land, and the oceans. Although models have been able to encompass more and more of the complexity found in the real world as computing power has increased, the still missing complexity will always leave a gap between 'modeled outcomes' and 'real outcomes'.

These uncertainties explain much of the range in forecasted climate change by 2100, from an increase of as 'little' as 2°C to as much as 8°C in average global temperatures. Notwithstanding these uncertainties, climate modelers and policy-makers give today's GCMs significantly better grades for their likely ability to forecast future global climate. Improved forecasting of future global climate, however, does not necessarily translate into being able to forecast the localized impacts of global climate change. Localized impacts will determine corporate exposure to, and vulnerability to, climate change hazards including wind conditions, temperature extremes, storm surges, and any of many other potential localized manifestations of climate change, but the spatial resolution of most GCMs is still far too coarse for such localized forecasting.

In light of these issues, companies can legitimately question whether they can really anticipate the potential impacts of climate change on their operations and supply chains in a way that supports useful adaptation planning.

## Addressing forecasting uncertainty

International development banks, among others, increasingly use the term 'climate proofing' to reflect their goal for climate change adaptation efforts involving a piece of infrastructure or a particular development project. The notion of 'climate proofing' is intuitively attractive, and a lot of the growing business and marketing literature around climate change adaptation seems to suggest that it is eminently doable.

Precisely or even accurately forecasting localized future climate conditions for adaptation planning, however, faces challenges that are unlikely to be overcome anytime soon, even as GCMs continue to improve. So how are forecasting uncertainties being addressed in current adaptation efforts? Three approaches characterize much of today's adaptation planning work:

- *Simplified model downscaling.* Because most GCMs model the climate at relatively coarse resolutions, e.g. 20 x 20 km and larger grids, generating higher resolution forecasts of climate change impacts requires that the results of the GCMs being 'downscaled'. Downscaling can in principle get you to a much finer resolution, including as low as 1 x 1 km grids (which is 400 times as detailed as a 20 x 20 km grid). There are two primary downscaling techniques, dynamic and statistical, each with its own advantages and disadvantages. Statistical downscaling is more commonly used in adaptation analysis today because it can be much simpler. For example, one can estimate the relationship between a localized climate variable (e.g. temperature on a specific mountaintop) and that same variable at the level of the GCM grid within which that mountain occurs (e.g. temperature averaged over the 400 sq. km area). When the GCM is run to reflect a climate change scenario it generates a changed temperature for the 400 km sq. area. That change in temperature can then be 'downscaled' into a forecasted changed temperature on the specific mountain top simply by relying on the previously established relationship between today's temperatures at the two scales. The problem with such a simplistic approach is the uncertainty around whether the statistical relationship between today's temperatures at the two points will remain the same under the climate change scenario. This is by no means assured, but simple statistical downscaling assumes just that. Dynamic downscaling, on the other hand, tries to model localized conditions in as much detail as possible, in effect creating a localized climate model that is initially populated with the results of the Global Climate Model and then run to generate estimates of localized climate change. Dynamic downscaling therefore requires a great deal more work and information than statistical downscaling as it is often applied today.

- *Changed decision-making methodologies.* One decision-support methodology developed with the goal of supporting adaptation planning is Robust Decision-Making (RDM). RDM asserts that one can arrive at robust adaptation decisions regardless of uncertainty about climate change forecasts. One approach modelers use in RDM is to ignore the probability distribution of climate outcomes, thereby avoiding the need for sophisticated modeling, and instead to try and come up with adaptation decisions that perform well regardless of future outcomes. It is not clear how a decision-making framework that ignores the best information regarding future conditions will lead to a robust answer. Any decision that performs well regardless of outcome cannot be taking advantage of the best available information, and must reflect an absolute least-common denominator approach to adaptation planning.

- *Building 'optionality' into adaptation decision-making.* As cities and other organizations start to look at major investments in climate change adaptation, the uncertainties of the prior two approaches become more obvious. When approaching decisions like how to upgrade the Thames Barrier to protect London from sea level rise, billions of dollars are at stake. To decide today how to upgrade the barrier for conditions that will not take place for decades could be considered financially imprudent, as would a decision to upgrade the barrier on the basis of assuming that all potential outcomes are equally probable. Instead, the River Thames project has developed a sophisticated approach to maintaining the 'optionality' of the process. In coming decades, for example, planners will learn more about the actual levels of sea level rise affecting the River Thames and London. This gives decision-makers time to adjust their strategy to better information before investing very large sums of money, as long as the infrastructure decisions they do make today create a foundation for flexibility, rather than locking in 'mal-adaptation' through faulty assumptions about the future. Making near-term decisions that don't account for the future, and which don't build in future response flexibility, could result in significant stranded investments or far more expensive future retrofits.

# Decision-making uncertainty in adaptation planning

We've just explored how several adaptation planning processes currently deal with uncertainty in climate forecasting. Efforts at adaptation planning can also encounter barriers based on the realities of corporate decision-making under uncertainty.

In principle, a climate change adaptation decision, at least for particular piece of infrastructure, could seem analytically and quantitatively rigorous, with a clear outcome. Take the designing of an offshore oil rig with a planned 100-year life. In an idealized planning situation, the following process might be employed:

- Step 1: appropriate global climate models would be combined into a model ensemble that would be downscaled and combined with regional climate models and environmental models (e.g. wind and wave models) to generate the needed climate change indicators for the structure, e.g. forecasted changes in wind and wave conditions.

- Step 2: based on the estimated probability distribution for climate change-induced wind and wave events, the vulnerability of the rig to climate change would be calculated and monetized (e.g. as a net present financial value of potential future damages).

- Step 3: engineering studies would assess potential adaptation measures, e.g. reinforcing the rig or raising the deck above forecasted wave heights.

- Step 4: economic analysis would identify the cost-effectiveness of alternative adaptation strategies.

- Step 5: using cost–benefit analysis, the right adaptation measures would be incorporated into the design of the rig.

This is an idealized adaptation example that involves just one piece of infrastructure facing just a couple of climate change hazards (wind and waves). Even then, modeling results are probabilistic; we can't predict with certainty what the winds and waves will be over 100 years. But the results of all of the steps above can be made to look quantitatively quite precise, especially if uncertainty bands are left out of the analysis. In reality, however, all the steps above involve estimating and managing probabilities, and as you layer on more and more uncertainty, the uncertainty bands are getting bigger.

With enough effort a company can reliably assess its exposure and its vulnerability to a given level of climate hazard. So if we knew future wind and wave conditions where this rig will be placed, it would be straightforward to prioritize the risks, identify risk reduction options, and select a robust adaptation strategy. Unfortunately, no amount of hard work can entirely eliminate the uncertainty around hazard estimation. In the case of the rig, you will inevitably be faced with a probability distribution of potential outcomes for each hazard reviewed (e.g. wind and waves). This will then filter through into probability distributions regarding exposure of the platform to the hazards, and then into probability distributions of the platform's vulnerability to the hazards. At the end of the day, the risks facing the platform will be in the form of a probability distribution.

Researchers at Det Norske Veritas carried out the idealized analysis described above for an offshore oil rig to better understand how climate change forecasts can contribute to corporate adaptation decision-making. The results are instructive in anticipating the challenges that companies will encounter in pursuing climate change adaptation strategies. The results suggested that given forecasted climate change, it would be economically worthwhile to raise the rig by 1–3 meters to avoid 'wave in deck' events during its 100-year lifetime.[30] But it wasn't actually that simple, and a real adaptation decision-making process would likely have encountered several hurdles:

- *Model uncertainty.* The GCMs and other models being used to forecast wind and wave conditions delivered substantially different results, with two out of three suggesting only a minimal risk of 'wave in deck' events, and the third suggesting a higher level of risk. What is the appropriate adaptation response? To treat the models as an ensemble and average the three very different results? This could lead to an under-designed rig if the more extreme result turns out to be more accurate, or an over-designed rig in the opposite case. Or perhaps, as is often done in such planning, the 'outlier' modeling result could be discounted and ignored, which in this case would lead to few if any changes to the rig's design.

- *Risk significance of the results.* Oil rigs are already designed to be resilient to pretty extreme events. In this case the models suggested that the probability of a 'wave in deck' event might go from 1 in 4000 years under today's climate conditions, to 1 in 600 years under future climate conditions. Even if adapting the design of the rig to this changed risk profile pencils out as cost-effective, decision-makers may question spending millions of extra dollars to address a 1-in-600-year event, especially in light of the model uncertainties already mentioned.

- *Discount rates.* The cost-effectiveness of raising the oil rig by 1–3 meters was based on net present value calculations using a 3% discount rate. Commonly termed a societal discount rate, this is a much lower figure than is commonly used in corporate planning, where 9–15% discount rates prevail. Higher discount rates, however, rapidly shorten the 'risk timeframe' of analysis. At a 15% discount rate, you don't care too much what happens after little more than a decade, even if a piece of infrastructure is supposed to last for 100 years.

This simple case study suggests that adaptation planning, even if rigorously pursued, will often not lead to obvious and economically unambiguous conclusions. Adaptation planning will have to be characterized by decision-making under uncertainty (and sometimes quite substantial uncertainty).

..........................................................................................

# CHAPTER 5

# An 'Adaptive' Approach to Managing Physical Climate Risks

**STANDARD BUSINESS PRACTICE** for many companies will soon require a climate change adaptation strategy of some kind. Companies could stick to 'adapting to the weather', but why do that if there's a better risk management option available? If preparing for today's changing weather conditions is perceived as good corporate risk management planning, and if decision-makers take the time to consider the question of 'is it worth it' to expand their thinking to future climate change, many corporate executives will answer the question in the affirmative when presented with the best available information.

The bigger challenge for corporate decision-makers is probably the question of 'can I do it?' The uncertainties associated with forecasting climate change on a timeframe and at a scale that is relevant to corporate decision-makers can appear daunting. As previously noted, a key part of answering the 'can I do it' question is to properly frame what 'it' is. An adaptation strategy that is expected to 'climate proof' a company for decades into the future will be exceedingly rare, no matter how many resources might be dedicated to the effort.

Searching for a perfect adaptation strategy is not the right way to frame the 'can I do it' question. Instead, what 'it' really refers to is reducing business risk by prudently expanding the horizon of corporate risk management from 'weather' to 'climate change'. Framed this way, it is easier to explain to decision-makers how a company can reduce the current uncertainty around the climate hazards it faces, better understand their exposure and vulnerability to such hazards, and be in a better position to quantify climate risk. Over time, incorporating new information as it becomes available, and going through further evaluations of exposure and vulnerability, should lead many companies to be able to significantly reduce their overall climate risk profile.

## Adapt to the weather and to climate change

As already noted, developing a climate change adaptation strategy does not mean having to 'climate proof' a company's operations against conditions that might not appear for decades into the future. But it does suggest that near-term decisions consider future climate risk in order to minimize future stranded investments, and to hedge against potentially accelerating climate change. Expanding one's perspective to climate change should generate several benefits:

- The massive amount of 'noise' associated with today's weather patterns complicates the process of figuring out what you are adapting to; clear conclusions can be hard to find. Taking a longer-term view as part of a climate change adaptation strategy can help in interpreting near-term trends as well.

- Adapting to the weather, done at the local level by operational staff, can result in a patchwork of responses to changing climatic conditions that confuses stakeholders and complicates corporate-wide investment decisions. A coherent climate change adaptation response will be based on corporate-wide risk management principles, helping avoid such confusion.

- Focusing only on 'the weather' may leave game-changing risks and opportunities unidentified. What if a company's entire business model is at risk over the next 20–40 years due to 'high confidence' climate forecasts? Shouldn't that information find its way to the C-Suite before it finds its way to investors, allowing the company to consider and manage both climate and brand risks?

When decision-makers suggest that they're doing as much as they can, and can't go further in addressing climate risks, they are implicitly answering either the 'is it worth it' or 'can I do it' questions in the negative. For companies with long-term assets, or with vulnerable operations and supply chains, understanding and managing climate change risks is prudent enterprise risk management. Many companies are likely failing this prudency test, simply because they're not asking the right risk management questions.

If a company does decide to expand its risk management horizons from today's weather to future climate change, the remaining question really is: how can a company develop an approach to reducing business uncertainty around future climate change that is both practical and cost-effective in the short term, and which continues to learn over time from better information that can further reduce uncertainty.

In *The Signal and the Noise* (Penguin, 2012), Nate Silver explores decision-making under uncertainty across topics from baseball to politics and climate change. Using Bayes' Theorem, Silver explores iterative learning processes that make maximum use of available information, while incorporating new information over time in order to reduce remaining uncertainty. The approach to uncertainty management Silver describes is also a good fit for many climate change adaptation planning needs, as described in the following six steps.

1. *Assess corporate vulnerability* to climate change hazards, emphasizing the need to take a 'broad brush' approach to quantifying business vulnerabilities, focusing on those aspects of operations, investments, and supply chains where vulnerability is most evident. This step should identify the potential manifestations of weather and climate change that are most relevant to a company's business model. Is it extremes that disrupt corporate operations or distribution systems, extremes that impact important nodes in a corporate supply chain, or other impacts?

2. *Understand what, if anything, is already happening.* Once important vulnerabilities are identified, compare recent and more historic trends in those variables to see whether relevant conditions are already changing. For example, comparing relevant weather extremes for the last 20 years to the extremes over the last 50–100 years can generate considerable insight into the ways in which the weather is changing, and the pace of such change.

3. *Assess corporate exposure* to climate change hazards forecasted to occur over a business-relevant timeframe, casting a wide net over corporate operations, investment decisions, and supply chains. If anything more than 30 years out is categorically beyond the decision-making relevance of anything about which a company is concerned, then climate impacts unlikely to occur sooner than 30 years in the future can be excluded from the assessment. For example, sea level

rise is one potential climate hazard, but we can be pretty confident that sea levels will not rise 3 feet in the next 30 years.

4. *Revisit corporate vulnerability* to climate change hazards based on the earlier steps, again emphasizing the need to take a 'broad brush' approach to quantifying business vulnerabilities, focusing on those aspects of operations, investments, and supply chains where exposure was judged to be real.

5. *Structure business risk hypotheses* around climate hazards for which vulnerability is judged to be potentially material. A company may conclude its vulnerability to changes in extreme summer temperatures could be business-material given the nature of its work and workforce. Based on its current understanding of the extreme temperature hazard, and its exposure and associated vulnerability to that hazard, company planners might settle on one of several hypotheses for adaptation strategy purposes:

   a. *Hypothesis 1.* Changes in extreme temperatures are highly unlikely to materially affect our operations in a timeframe relevant to our corporate planning, and therefore the hazard does not merit pro-active adaptation measures.

   b. *Hypothesis 2.* We are already seeing changes in extreme temperatures, and forecasted extremes could have a material impact in a timeframe relevant to our corporate planning. We should track the topic and develop contingency plans for dealing with that vulnerability.

   c. *Hypothesis 3.* We are already vulnerable to observed temperature extremes, and we forecast the level of vulnerability to increase significantly within 10 years. We need to move aggressively to develop measures by which to mitigate either our exposure or our vulnerability to these extremes to keep risk within acceptable levels.

6. *Update hypotheses as well as exposure and vulnerability estimates.* The beauty of the Bayesian approach to reducing uncertainty over time is that a company doesn't need to get too hung up about knowing or predicting the future at the beginning of the process. Rather, the company can focus initial effort where the organization has the most confidence in its analysis, namely assessing exposure and vulnerability to alternative hazard outcomes. The associated hazard hypotheses can start from almost any point, whether suggesting almost no risk, to suggesting very material risk. If the chosen hypothesis is significantly off base it will likely become clear relatively quickly as the hypothesis is tested against new information. The hypothesis, and associated adaptation conclusions, can then be appropriately modified.

Step 6 above is critical to this process; without the feedback and learning loop, risk reductions cannot be assured. Companies need to regularly revisit whether they observe material impacts, or whether best available information suggests changes to previous assessment of the likelihood of such material impacts. A Bayesian approach to learning from new information allows initial hypotheses regarding the business materiality of climate impacts to be regularly updated based on improving knowledge and science, and risk management strategies to be adjusted accordingly.

The Bayesian-inspired approach provides a practical means for developing corporate adaptation strategies, and is already reflected in

*The climate change sensitivity framework being developed by Rio Tinto Alcan is not intended to predict future climate change by quantifying and reducing the uncertainty of projections. Instead, it accepts that some uncertainties associated with projected climate change are irreducible, and takes account of a range of potential future greenhouse gas emissions scenarios and global climate models. The framework depends on the expert input of Rio Tinto staff, and emphasizes learning from past events. The framework also incorporates a risk matrix that highlights risks that can be addressed as a matter of priority. Instead of a top-down methodology that attempts to foresee the future, Rio Tinto Alcan is building a bottom-up approach that increases the group's capacity to deal with the unexpected.*[31]

some companies' approaches to climate change adaptation.

The approach suggested here does not mean that companies should necessarily prioritize climate risks differently from other risks, or change the risk management criteria they apply to climate risks as compared to other risks. We do believe, however, that if companies applied the same risk management approaches to climate change risks that they routinely apply to other risks, they would internalize climate risks more actively into their enterprise risk management strategies.

Companies should also consider higher-level societal considerations, such as corporate social responsibility programs or sustainability commitments, in evaluating adaptation strategies. We do not mean to suggest that such initiatives are irrelevant. We simply assume that company strategies dealing with climate risk and adaptation to climate change can reasonably be expected to look out for the interests of the company, shareholders, and other stakeholders, and this fact of life is not likely to change anytime soon.

# The importance of executive leadership

Climate change adaptation planning can appear a technical and analytical topic of inquiry. Rarely will the results of adaptation analysis point in a completely unambiguous and cost-effective direction. Effectively incorporating climate change adaptation objectives into corporate risk management strategies will usually require executive leadership or those efforts will founder. In the absence of leadership, addressing climate change adaptation could become a pro-forma exercise in which the 'adaptation box' is checked off on infrastructure and other projects, without anyone having looked at the issue seriously in the context of risk management.

Executive guidance will be needed in several areas:

- *Thinking outside the box.* The many relationships and dependencies create a challenging aspect of anticipating impacts of climate change on a company's operations and supply chains. Until the company takes a holistic look, with an explicit effort to think 'broad brush' about what those impacts might be, many companies will likely characterize impacts as 'unknown unknowns' that in retrospect really weren't 'unknown'. Adaptation thinking can't safely be buried within an organization and left up to exactly the same processes and engineering guidelines used today in designing infrastructure and projects.

- *Timeframe.* If the timeframe for looking at climate change is too short, it defeats the purpose of trying to anticipate risks related to climate change. Companies should strive to institutionalize as long-term a thinking process as can be justified by the types of decisions being made (e.g. long-term infrastructure investments), the nature of the business itself, and the company's sustainability commitments, among other variables. Without executive leadership, the short-term interests of individual departments are likely to counter this need.

- *Risk aversion.* Companies find it difficult to deal with uncertain outcomes and the probability distributions associated with those outcomes; thus, it is common to focus on the 'most likely' outcome. In the case of climate change impacts, this may leave dramatic risks on table, buried in the 'long tail' of the risk distribution. Should companies ignore a 10–30% risk of outcomes that could lead to critical damage to the corporation, in favor of focusing on the assumed 50% likelihood that the impacts will be more modest? In the face of accelerating climate change, this could quickly turn into a high-risk assumption that companies will wish they hadn't made. Executive leadership is needed to frame the degree of risk aversion to bring to the adaptation planning process.

- *Discount rate.* How does the company discount the future value of money and implicitly future risks? For a subject like climate change, where so many business and societal interests are at stake, decision-makers should carefully evaluate how to value long-term risks and returns.

- *Public policy.* As we argued in *The Changing Nature of Corporate Climate Change Risk*, companies have tended to focus on public policy as the primary source of climate risk. Decision-makers should recognize that the two are linked. In the face of substantial climate change, for example, we could trigger what we've previously termed the Climate Response Tipping Point, the point at which climate politics fundamentally shift from 'avoidance mode' into 'response mode'. Changing climate outcomes could easily influence climate policy outcomes, potentially presaging a climate risk 'perfect storm' for many companies.

.......................................................................................

# CHAPTER 6

# Conclusions

**BLOOMBERG BUSINESSWEEK CHARACTERIZED** the prevailing perception that 2012 signaled a substantial shift towards more extreme events with its dramatic cover story (see Figure 8). 2012 was indeed an extraordinary year for weather and climate. It was the warmest year ever recorded in the United States[32] and the 10th warmest year globally.[33] In the Arctic, 2012 set multiple sea ice records, including a minimum ice extent 18% below the previous record set in 2007, and 49% below the 1979–2000 average.[34]

**FIGURE 8. Front cover of *Bloomberg Businessweek*, 1 November 2012.**

In the United States, 2012 was the year that a low-level Category 1 Hurricane became Superstorm Sandy, accompanied by a 13.88 feet storm surge that broke the all-time record in New York Harbor, causing large-scale damage in the City and in surrounding states. The US witnessed 362 all-time high temperatures in 2012, and no all-time lows.[35] The US also suffered the worst drought in 50 years, with over 1300 US counties across 29 states declared drought disaster areas, and important rivers reached record-low flows, impeding river traffic.[36] Meantime, England had its wettest year ever,[37] and Nigeria suffered from record floods that killed hundreds and displaced millions.[38]

Whether the world will look back and see that 2012 marked a transition in societal and business thinking about climate change adaptation cannot yet be known. What is clear is that the nature of the discussion over climate change and its impacts has changed. The prominence of climate change adaptation in the consciousness of the policy community, the public, and the business community is much higher today than even a year ago. The business community will face more policy and stakeholder interest in what they are doing to adapt to climate change, a subject most executives have had relatively little experience with to date.

What corporate decision-makers need to do in terms of climate change adaptation differs markedly from how they have managed mitigation efforts for years. Climate change adaptation cannot simply be turned over to the marketing department, or rolled into the responsibilities of the Chief Sustainability Officer. Adaptation is local, it is risk-based, it is operational, and it is strategic.

Executives may find that individuals in their companies have already been doing a lot of thinking about adapting 'to the weather' (see Jeffrey Williams's quote below). This thinking can be a very useful foundation, and if strengthened and expanded to corporate adaptation to climate change, lead to a much more cohesive risk management strategy overall.

> *"You cannot manage a risk if you deny it exists, or don't see it coming."* [39]
>
> **Jeffrey Williams**
> Director, Climate Consulting,
> Entergy Corporation

Adaptation to climate change will inevitably become an element of corporate enterprise risk management, corporate environmental policy, and business prudence for many companies either exposed to or vulnerable to such impacts. The sooner corporate leaders start to prepare their organizations for this shift, the easier the process. Corporate decision-makers will increasingly conclude both that adaptation is 'worth it' from a corporate risk management perspective, and that they can successfully integrate adaptation in corporate risk management processes. Corporate decision-makers need to understand that they have an important leadership role to play in structuring the process.

# Notes

1. http://www.ipieca.org/news/20121016/addressing-adaptation-oil-and-gas-industry-workshop-held.

2. Patterson, K., et al. 2008. *Influencer: The Power to Change Anything* (New York: McGraw-Hill Publishing).

3. Trexler, M. and Kosloff, L., 2012. *Understanding and Managing the Changing Profile of Corporate Climate Change Risk* (Oxford: DōSustainability), at p. 43.

4. http://www.ipcc.ch/pdf/glossary/ar4-wg3.pdf.

5. Grossman, D. 2012. *Physical Risks from Climate Change* (Calvert Investments, Ceres, and Oxfam America), at p. 4.

6. National Round Table on the Environment and the Economy. 2012. *Managing the Business Risks and Opportunities of a Changing Climate: A Primer for Executives on Adaptation to Climate Change* (Ottawa: National Round Table on the Environment), at p. 5.

7. http://blogs.windsorstar.com/2012/08/09/city-has-a-draft-plan-for-climate-change/.

8. The following examples are all taken from Grossman (2012).

9. Amado, J.C. and Adams, P. 2012. *Value Chain Climate Resilience: A Guide to Managing Climate Impacts in Companies and Communities* (Montreal: Acclimatise), at p. 3.

10. Ridley, M. and Duguid, B., 2009. Climate change – a business revolution? Carbon Trust presentation, London, slide 37, available at: http://www.scottish oilclub.org.uk/lib/Presentation_090423_MichaelRidley_CarbonTrust_ScottishOilClub.pdf.

11. Backus, G. et al. 2010. *Assessing the Near-Term Risk of Climate Uncertainty: Interdependencies among the U.S. States* (Albuquerque, NM: Sandia Corporation).

12. World Bank. 2012. *Turn Down the Heat: Why a 4 Degree Centigrade Warmer World Must Be Avoided* (Washington, DC: International Bank for Reconstruction and Development/The World Bank).

13. National Round Table on the Environment and the Economy, 2012. *Facing the Elements: Building Business Resilience* (Ottawa: National Round Table), at p. 19.

14. Backus et al. (2010), at p. 24.

15. http://www.cfr.org/energy/new-north-american-energy-paradigm-reshaping-future/p28627.

16. http://articles.chicagotribune.com/2012-08-01/business/sns-rt-usa-environ mentnorthcarolinal2e8j1kfe-20120801_1_coastal-resources-commission-coastal-economic-development-group-coastline-and-thousands.

17. National Round Table on the Environment and the Economy (2012), at p. 54.

18. National Round Table on the Environment and the Economy (2012), at p. 82.

19. National Round Table on the Environment and the Economy, at p. 59.

20. Entergy, 2011. *Building a Resilient Energy Gulf Coast.*

21. Sathaye, J. et al. 2012. Estimating risk to California energy infrastructure from projected climate change. Presentation to Atlantic Council workshop, July 2012.

22. National Round Table on the Environment and the Economy (2012), at p. 34.

23. National Round Table on the Environment and the Economy (2012), at p. 3.

24. Kahneman, D. 2011. *Thinking Fast and Slow* (New York: Farrar, Straus & Giroux).

25. Nick Mabey, personal communication. February 2013.

26. http://stanfordhospital.org/newsEvents/newsReleases/2011/vderm.html.

27. Amado and Adams (2012), at p. 4.

28. Amado and Adams (2012), at p. 7.

29. Association of Climate Change Officers (ACCO), 2011. *Barriers to Climate Change Adaptation Plans: A Survey of Climate Professionals* (Washington, DC: ACCO).

30. Garre, L., Friis-Hansen, P. and Trexler, M. 2012. Adapting engineered structures to climate change. Presentation to Nordic International Conference on Climate Change Adaptation, 29–31 August 2012, Helsinki, Finland. Available at: **http://www.nordicadaptation2012.net/Doc/Oral_presentations/5.4.2_Trexler.pdf**.

31. National Round Table on the Environment and the Economy, 2012. *Facing the Elements: Building Business Resilience in a Changing Climate* (Ottawa, Canada), at p. 62

32. **http://www.npr.org/blogs/thetwo-way/2013/01/08/168882644/**its-in-the-books-2012-was-warmest-year-on-record-for-lower-48-states.

33. **http://www.businessweek.com/news/2013-01-15/**2012-was-world-s-10th-warmest-year-on-record-noaa-reports.

34. **http://www.tiki-toki.com/timeline/entry/55279/**Extreme-Weather-Climate-Events-2012/#vars!panel=**683139**!

35. **http://www.usatoday.com/story/weather/2013/01/08/**record-warm-year-**2012/1817841/**.

36. **http://www.nrdc.org/health/extremeweather/**.

37. **http://www.tiki-toki.com/timeline/entry/55279/**Extreme-Weather-Climate-Events-2012/#vars!panel=**800799**!.

38. **http://www.tiki-toki.com/timeline/entry/55279/**Extreme-Weather-Climate-Events-2012/#vars!panel=**683133**!.

39. National Round Table on the Environment and the Economy, 2012. *Managing the Business Risks and Opportunities of a Changing Climate: A Primer for Executives on Adaptation to Climate Change*, Text box on p. 3.

..........................................................................................